MW01358059

This Book Belongs To

____

____

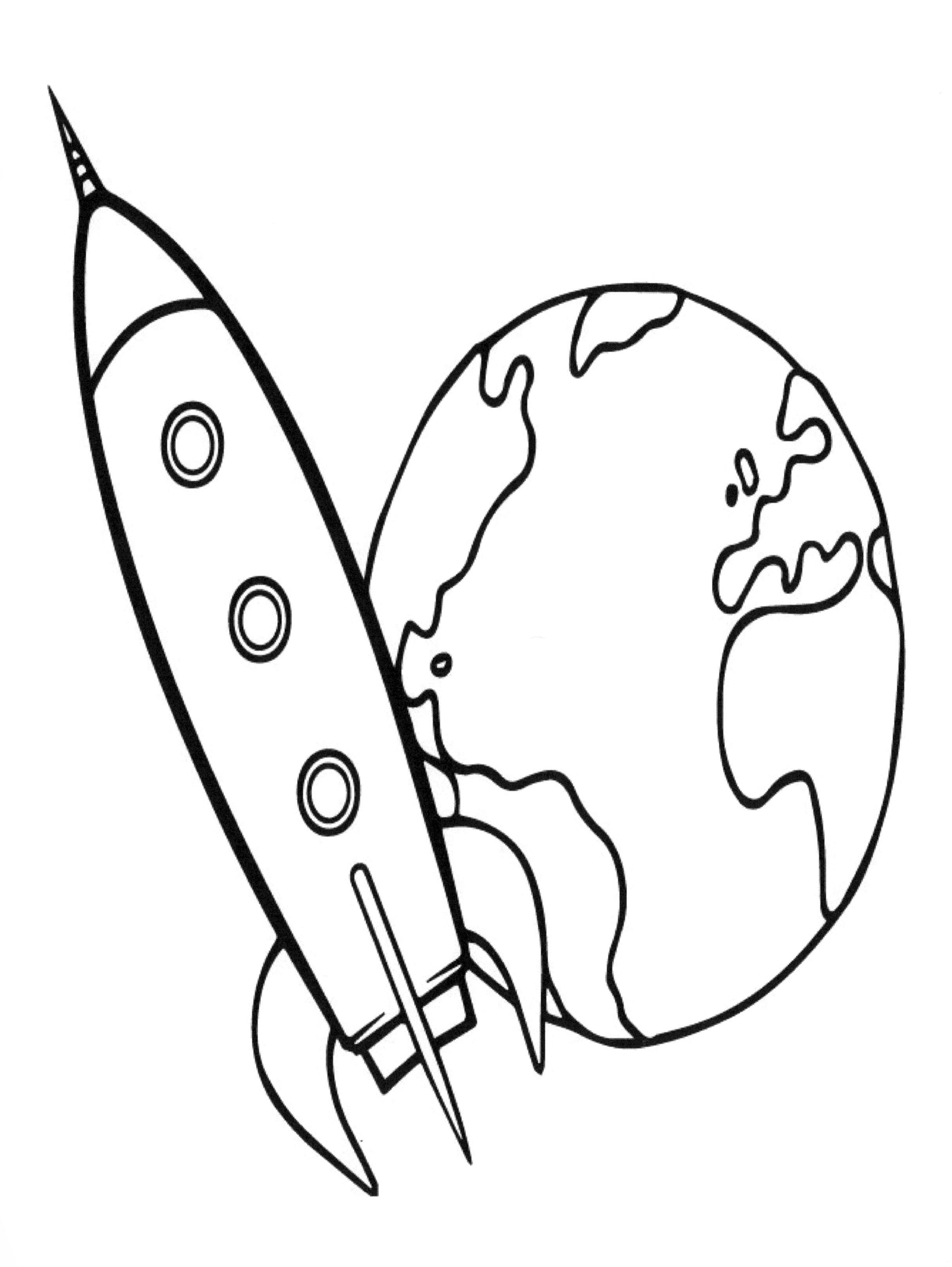

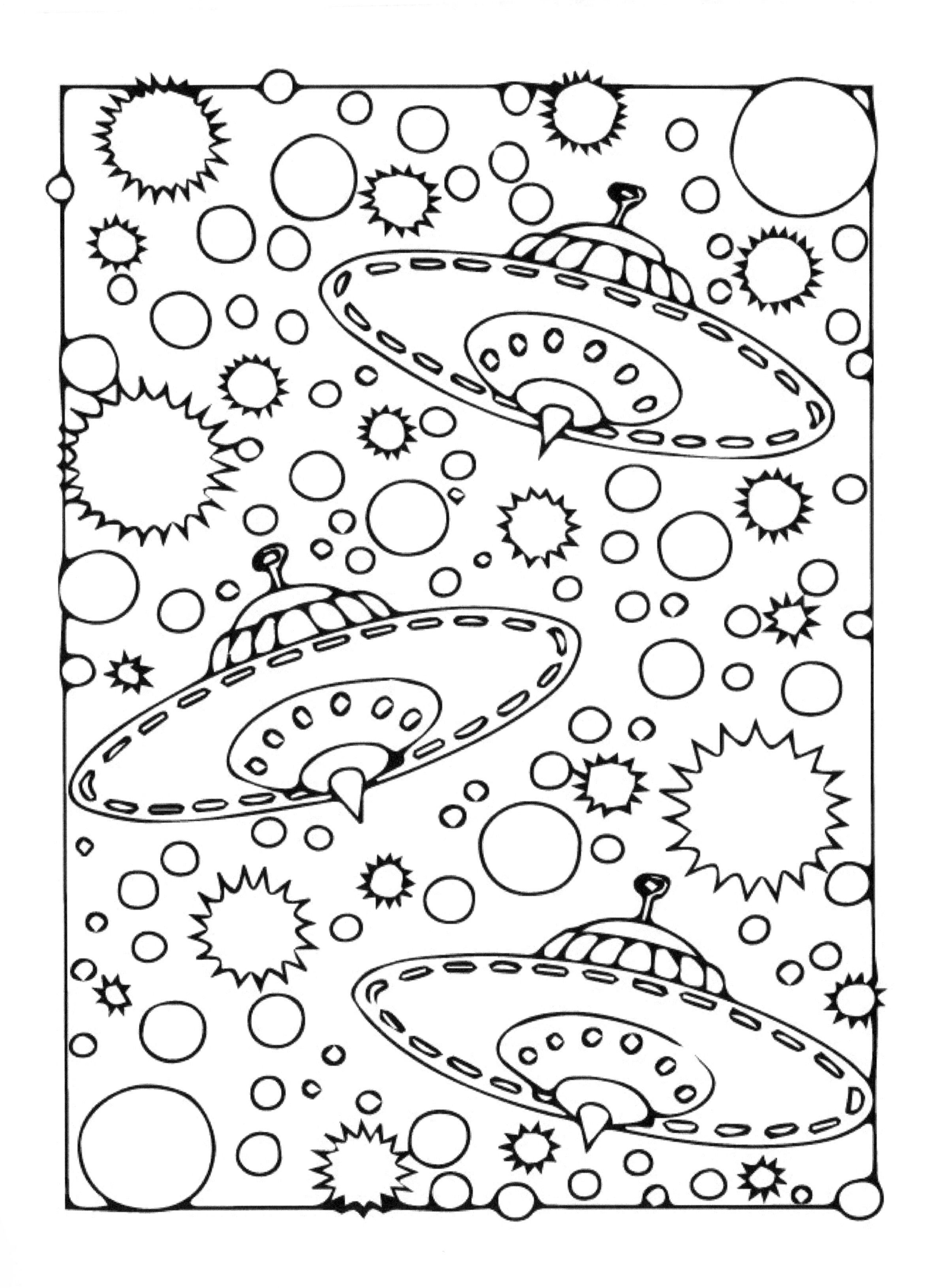

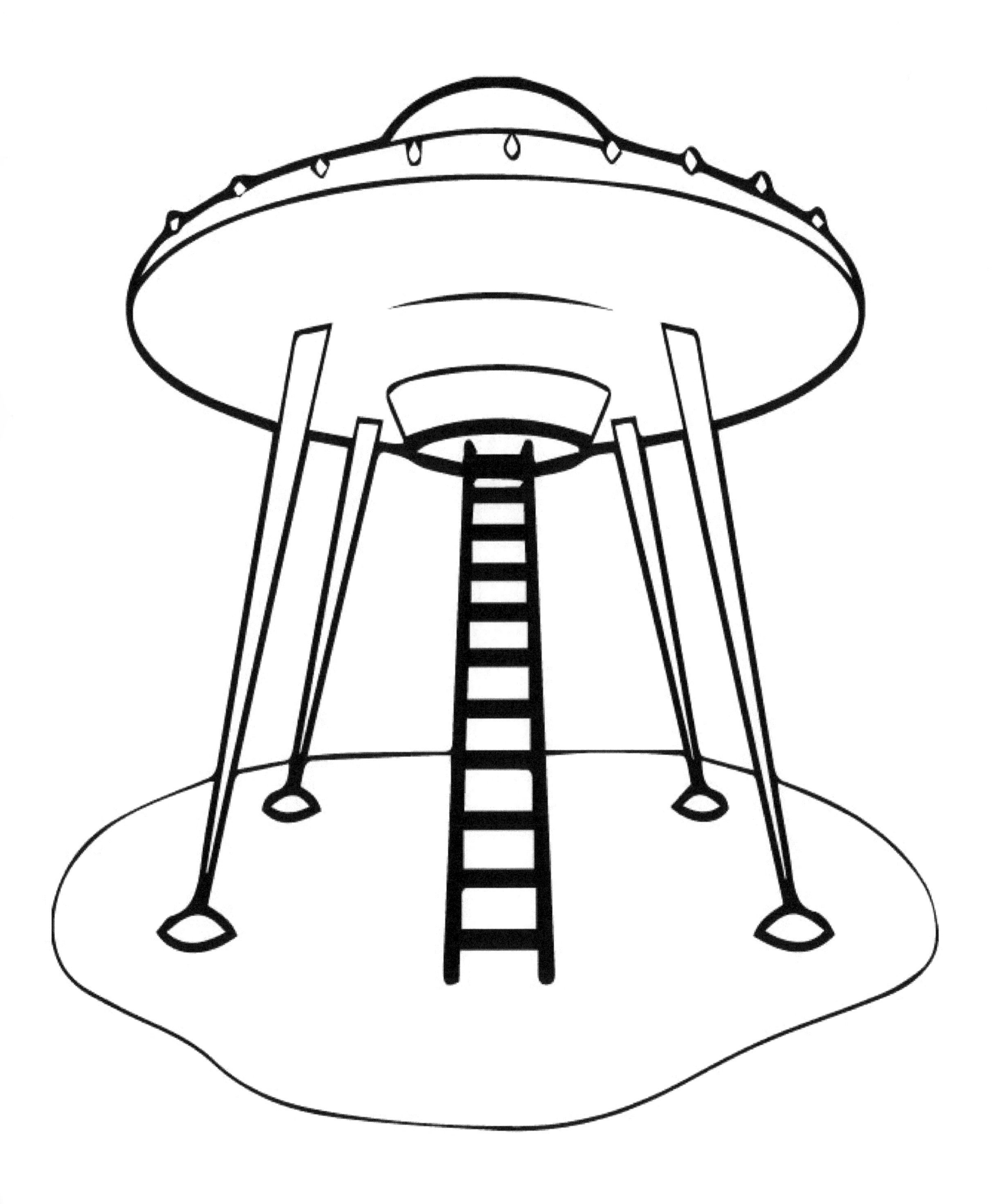

Mercury

Venus

Earth

Mars

Jupiter

Saturn

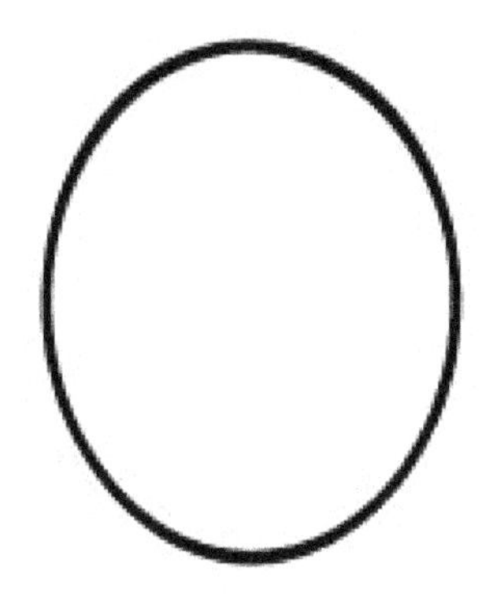

Uranus

Neptune

Pluto

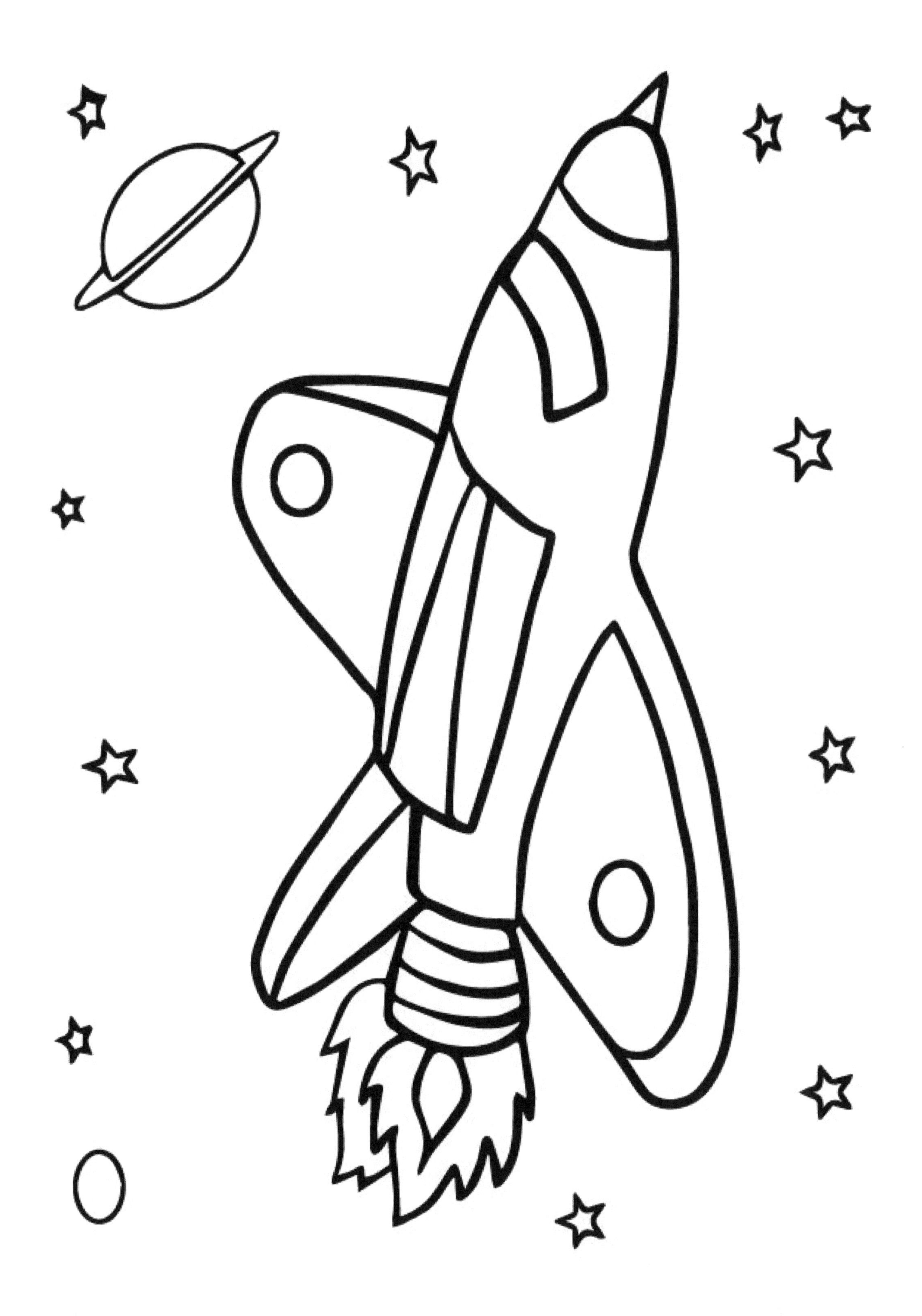

Printed in the USA
CPSIA information can be obtained
at www.ICGtesting.com
LVHW080909171124
796857LV00014B/662
9798611173817